恐龙俱乐部

鸟真的是恐龙吗？

Is It True That Birds Are Dinosaurs？

[英] 露丝·欧文/著

刘颖/译

汉英对照
恐龙科普

江苏凤凰美术出版社

全家阅读
小贴士

★ 每天空出大约10分钟来阅读。

★ 找个安静的地方坐下，集中注意力。关掉电视、音乐和手机。

★ 鼓励孩子们自己拿书和翻页。

★ 开始阅读前，先一起看看书里的图画，说说你们看到了什么。

★ 如果遇到不认识的单词，先问问孩子们首字母如何发音，再带着他们读完整句话。

★ 很多时候，通过首字母发音并听完整句话，孩子们就能猜出单词的意思。书里的图画也能起到提示的作用。

最重要的是，感受一起阅读的乐趣吧！

扫码听本书英文

Tips for Reading Together

• Set aside about 10 minutes each day for reading.

• Find a quiet place to sit with no distractions. Turn off the TV, music and screens.

• Encourage the child to hold the book and turn the pages.

• Before reading begins, look at the pictures together and talk about what you see.

• If the child gets stuck on a word, ask them what sound the first letter makes. Then, you read to the end of the sentence.

• Often by knowing the first sound and hearing the rest of the sentence, the child will be able to figure out the unknown word. Looking at the pictures can help, too.

Above all enjoy the time together and make reading fun!

Contents 目录

关键时刻
Crunch Time

一只饥饿的鸟用它锋利的牙齿咀嚼着蜻蜓。

这种鸟叫始祖鸟。

它生活在1.5亿年前的恐龙时代！

A hungry bird crunches a dragonfly with its sharp teeth.

The bird is called Archaeopteryx.

It lived 150 million years ago in the time of the dinosaurs!

始祖鸟英文名的字面意思是"古老的翅膀"。
The name Archaeopteryx means "old wing".

始祖鸟 Archaeopteryx
(ar-kee-OP-ter-ix)

长牙齿的鸟
A Bird with Teeth

科学家于1861年首次发现了始祖鸟化石。

它有一只大乌鸦那么大。

Scientists first found Archaeopteryx **fossils** in 1861.

It was the size of a large crow.

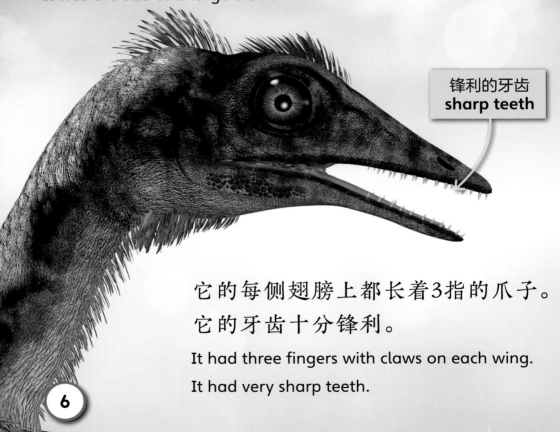

锋利的牙齿
sharp teeth

它的每侧翅膀上都长着3指的爪子。

它的牙齿十分锋利。

It had three fingers with claws on each wing.

It had very sharp teeth.

始祖鸟化石 Archaeopteryx fossil

带利指的长爪
fingers with claws

翅膀
wing

长长的骨质尾巴
long, bony tail

现在的鸟类有喙，但没有牙齿。
它们的翅膀上也没有爪子。

Today birds have beaks and no teeth.
They don't have claws on their wings.

鸟还是恐龙？
Bird or Dinosaur?

始祖鸟像恐龙，因为它有牙齿和爪子。
但它有羽毛，能像鸟一样飞翔。

Archaeopteryx was like a dinosaur because
it had teeth and claws.
But it had feathers and could fly like a bird.

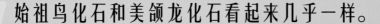

始祖鸟化石和美颌龙化石看起来几乎一样。

The fossils of Archaeopteryx and a dinosaur called Compsognathus look nearly the same.

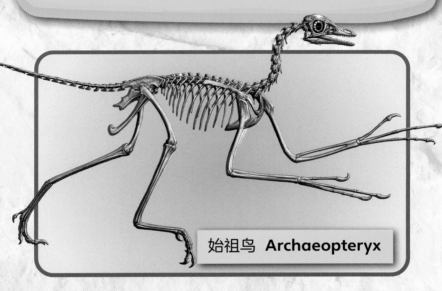

始祖鸟 **Archaeopteryx**

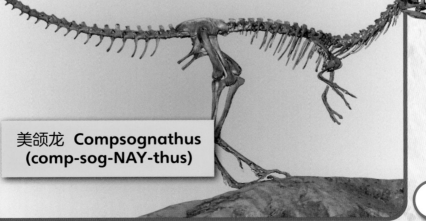

美颌龙 **Compsognathus
(comp-sog-NAY-thus)**

惊人的发现
An Amazing Discovery

科学家在研究过许多化石后，有了一些惊人的发现。鸟是从恐龙进化而来的！

Scientists looked at lots of fossils and found out something amazing. Birds **evolved** from dinosaurs!

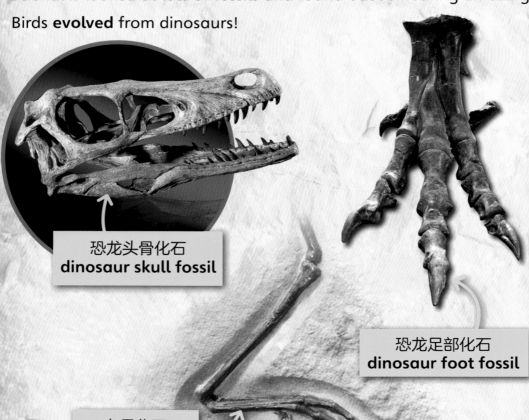

恐龙头骨化石
dinosaur skull fossil

恐龙足部化石
dinosaur foot fossil

鸟足化石
bird foot fossil

进化成鸟类的恐龙与迅猛龙属于同一个恐龙家族。
The dinosaurs that became birds were from the same dinosaur family as Velociraptors.

数千万年来，这些恐龙进化成了鸟类。
Over tens of millions of years, these dinosaurs evolved and became birds.

迅猛龙 Velociraptor
(vel-OS-ee-rap-tor)

许多鸟
Lots of Birds

有些史前鸟类会飞，

有些则不会飞。

Some prehistoric birds could fly.

Some birds were flightless.

巴塔哥尼鸟 Patagopteryx
(pata-GOP-ter-ix)

巴塔哥尼鸟是一种不会飞的鸟。

Patagopteryx were flightless birds.

燕鸟和鸡一样大。

它会飞，且以鱼类为食。

Yanornis was the size of a chicken.

It could fly and it ate fish.

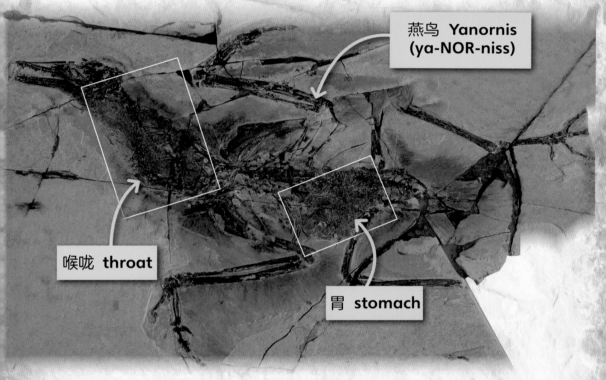

燕鸟 Yanornis
(ya-NOR-niss)

喉咙 throat

胃 stomach

科学家发现了一枚燕鸟化石，它的喉咙和胃里有鱼。

Scientists found a fossilized Yanornis with fish in its throat and stomach.

小翅膀
Tiny Wings

尾羽龙生活在中国。

它和孔雀一样大。

它长着小翅膀，但有翅难飞。

Caudipteryx lived in China.

It was the size of a peacock.

It had tiny wings and was flightless.

尾羽龙 **Caudipteryx**
(cow-DIP-ter-ix)

14

尾羽龙吃小石头来磨碎胃里的种子。
Caudipteryx ate small stones to grind up seeds in its stomach.

尾羽龙化石
Caudipteryx fossil

石头 **stones**

科学家在一具尾羽龙化石的胃里发现了石头。
Scientists found stones in the stomach of a fossilized Caudipteryx.

有喙的鸟
Birds with Beaks

孔子鸟和鸽子一样大，它会飞。

Confuciusornis was the size of a pigeon and it could fly.

孔子鸟 Confuciusornis
(con-FEW-shush-OR-nis)

它生活在1.2亿年前，但看上去像今天的鸟类。

它有喙，但没有牙齿。

它长着尾羽，而不是骨质尾巴。

It lived 120 million years ago, but it looked like birds we see today.

It had a beak, not teeth.

It had tail feathers, not a bony tail.

孔子鸟化石 Confuciusornis fossils

尾羽 tail feather

小行星撞击
Asteroid Crash

史前鸟类经历了什么？

大约6600万年前，一颗巨大的小行星撞上了地球。

What happened to the prehistoric birds?

About 66 million years ago, a giant **asteroid** crashed into Earth.

小行星 asteroid

撞击引发了火灾和洪水。
大多数史前鸟类因此死亡并灭绝。

The asteroid caused fires and floods.
Most of the prehistoric birds died and became **extinct**.

今天的鸟类
Birds of Today

一些史前鸟类在小行星撞击中幸存下来。
它们经历了数千万年的进化。

Some prehistoric birds survived the asteroid crash.
They evolved over tens of millions of years.

企鹅 penguin

它们变成了我们今天见到的鸟类。
They became the birds we see today.

所有鸟类都是从恐龙进化而来的。这就是为什么人们说鸟就是恐龙。
All birds have evolved from dinosaurs. That's why people say that birds are dinosaurs!

鹦鹉
parrot

巨嘴鸟
toucan

鸽子 **pigeon**

麻雀 **sparrow**

词汇表 Glossary

小行星　asteroid

宇宙中环绕太阳运动的岩石天体。
A large rock travelling through space around the Sun.

进化　evolved

在漫长的时间里一点一点地发生变化。
Changed bit by bit over a long period of time.

灭绝　extinct

永远消失。
Gone forever.

有翅难飞　**flightless**

无法飞行。

Unable to fly.

化石　**fossil**

存留在岩石中几百万年前
的动物和植物的遗体。

The rocky remains of an animal or
plant that lived millions of years ago.

史前　**prehistoric**

人类开始记录历史前的
一段时间。

A time before people began
recording history.

恐龙小测验 Dinosaur Quiz

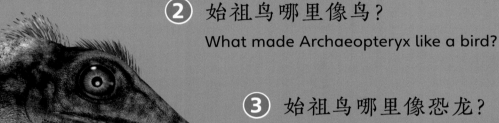

① 请说出始祖鸟和现代鸟类的一个区别。
Give one difference between Archaeopteryx
and a bird we see today.

② 始祖鸟哪里像鸟？
What made Archaeopteryx like a bird?

③ 始祖鸟哪里像恐龙？
What made Archaeopteryx
like a dinosaur?

④ 科学家是怎么知道燕鸟吃鱼的？
How did scientists know Yanornis ate fish?

⑤ 你认识多少种鸟？
How many different kinds of
birds do you know?